SD Gundam Battle Alliance

The Complete Guide & Walkthrough with Tips &Tricks

SD Gundam Battle Alliance: All Mobile Suits and How to Unlock Them

Looking for a list of All Mobile Suits in SD Gundam Battle Alliance? Well, you've come to the right place. For this guide, we've catalogued every suit that you can play as in the new action RPG, complete with details on how to actually unlock them.

To unlock a suit, you'll need to find Blueprints, which are detailed below. Some Blueprints are easy to find, while others may require a bit of forward thinking.

It's fair to say that there's an impressive range of suits to play as in SD Gundam Battle Alliance. Some are definitely more powerful than others, but overall, every suit can be effective in combat. You start out with a selection of relatively weak suits, but you'll soon build up your arsenal by simply playing through the game. This guide should help you keep track of the suits that you want.

On this page:

SD Gundam Battle Alliance: How to Unlock Mobile Suits

SD Gundam Battle Alliance: All Mobile Suits and How to Unlock Them

GM

Zaku II Type F

Guntank

Guncannon

Dom

RX-78-2 Gundam

Gundam Barbatos (6th Form)

GM Sniper II

Arios Gundam

Gundam Ground Type

Freedom Gundam

Gouf Custom

Seravee Gundam

Gundam Exia Repair II

Gundam Ez8

Reginlaze Julia (Final Battle)

Full Armor Gundam

Kampfer

Gundam "Alex"

Z'Gok (Char)

Providence Gundam

Zeta Gundam

Gundam Barbatos Lupus

Hyaku Shiki

Gundam Gusion Rebake Full City

Justice Gundam

Gelgoog (Char)

Gundam GP03S Stamen

Gundam F91

SD Gundam Battle Alliance: How to Unlock Mobile Suits

In SD Gundam Battle Alliance, you can take to the battlefield as all kinds of different mobile suits — but you'll have to unlock them first. You start the game with just a small handful of suits, but you'll quickly discover more over the course of the campaign. Generally speaking, each new mission will present a fresh opportunity to unlock a new suit.

Blueprints are needed to unlock each suit. Blueprints can be dropped by boss enemies, or they can be given to you as mission rewards. In some rare cases, Blueprints can also be found inside of specific containers hidden around levels.

Some suits will require just a single Blueprint to unlock, while others will require multiple. This means that you'll often have to replay missions in order to get the number of Blueprints that you need.

Fortunately, Blueprints are not dropped randomly. Each mission will always yield the same Blueprints, although you may sometimes get lucky and receive more than one for your trouble. Blueprints that you already own and have no use for are automatically sold for Capital, the in-game currency.

SD Gundam Battle Alliance: All Mobile Suits and How to Unlock Them

Below is a list of All Mobile Suits in SD Gundam Battle Alliance, with details on How to Unlock Them. This list is in chronological order, meaning that the suits at the top are discovered first, while the ones at the bottom are found much later in the game.

GM

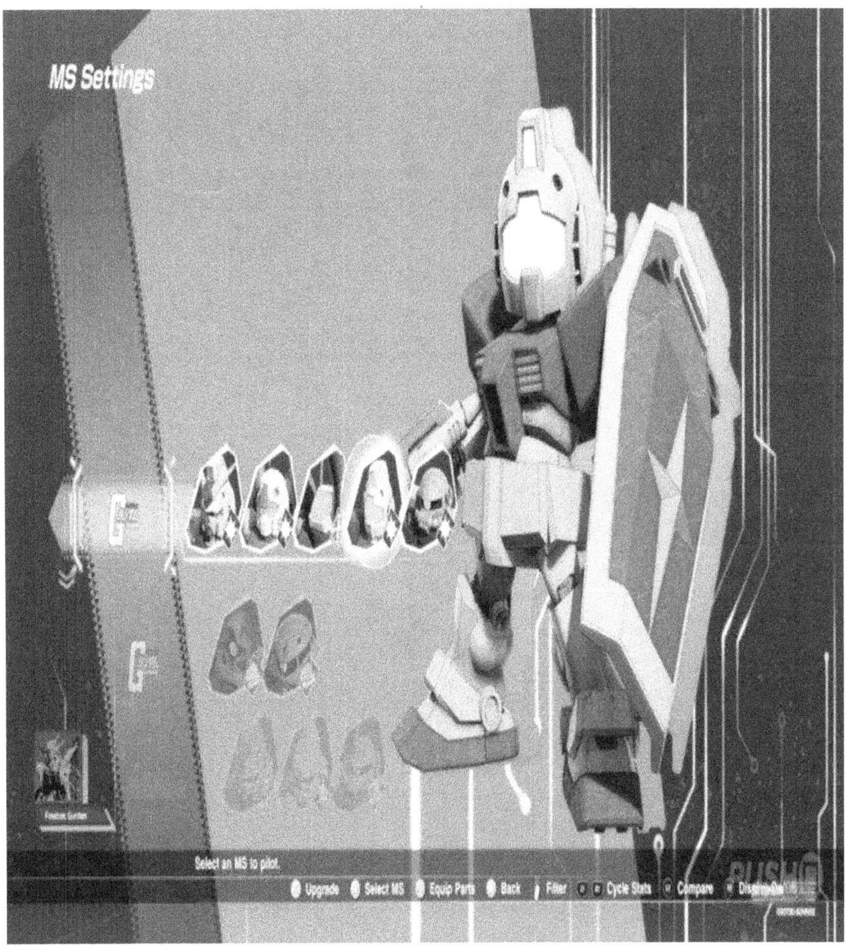

Gundam Series: Mobile Suit Gundam

Automatically unlocked after the opening mission.

Zaku II Type F

Gundam Series: Mobile Suit Gundam

Automatically unlocked after the opening mission.

Guntank

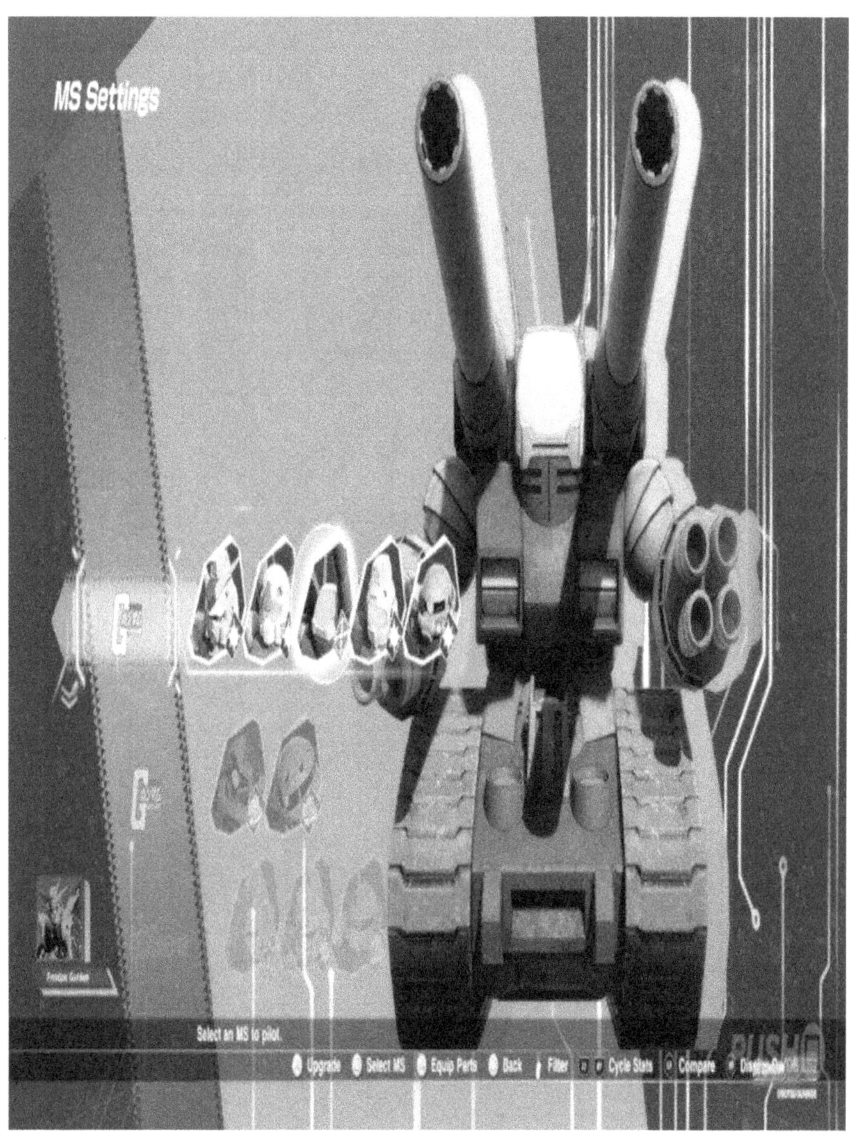

Gundam Series: Mobile Suit Gundam

Complete Break Mission: The Threat of Zeon (Directory 1)

Guncannon

Gundam Series: Mobile Suit Gundam

Complete Break Mission: The Threat of Zeon (Directory 1)

Dom

Gundam Series: Mobile Suit Gundam

Requirements:

Dom Blueprint I

Open the container next to White Base in True Mission: Ramba Ral's Attack! (Directory 1)

RX-78-2 Gundam

Gundam Series: Mobile Suit Gundam

Requirements:

2x Gundam Blueprint I

Dropped by the boss of True Mission: Ramba Ral's Attack! (Directory 1)

2x Gundam Blueprint II

Dropped by the boss of Irregular Mission: Trails to the Land (Directory 2)

3x Gundam Blueprint III

Dropped by the boss of True Mission: Big Zam's Last Stand (Directory 3)

Gundam Barbatos (6th Form)

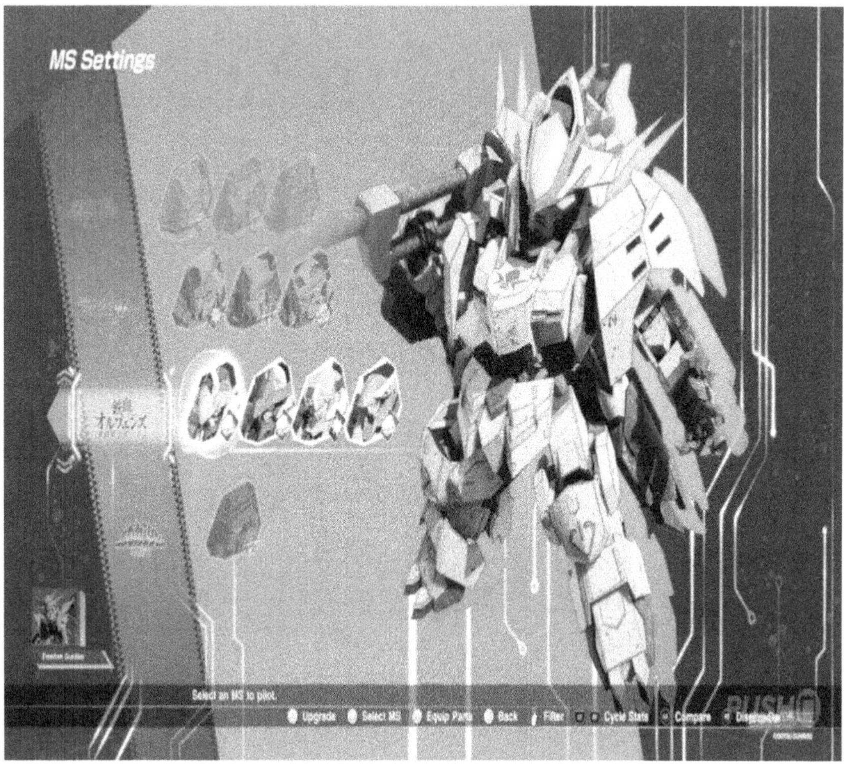

Gundam Series: Mobile Suit Gundam: Iron Blooded Orphans

Requirements:

2x Gundam Barbatos (6th Form) Blueprint I

Dropped by the boss of True Mission: Tekkadan (Directory 1)

GM Sniper II

Gundam Series: Mobile Suit Gundam 0080: War in the Pocket

Requirements:

2x GM Sniper II Blueprint I

Dropped by the boss of Irregular Mission: In the Name of G (Directory 1)

Dropped by the boss of Break Mission: Battle for Chryse (Directory 3)

Arios Gundam

Gundam Series: Mobile Suit Gundam 00

Requirements:

2x Arios Gundam Blueprint I

Dropped by the boss of Break Mission: Seen and Unseen (Directory 2)

Dropped by the second boss of True Mission: I Can Hear a Song (Directory 2)

Gundam Ground Type

Gundam Series: Mobile Suit Gundam: The 08th MS Team

Requirements:

1x Gundam Ground Type Blueprint I

Dropped by the boss of Break Mission: Seen and Unseen (Directory 2)

Freedom Gundam

Gundam Series: Mobile Suit Gundam SEED

Requirements:

3x Freedom Gundam Blueprint I

Dropped by the boss of Break Mission: Victory Song of the Resistance (Directory 2)

2x Freedom Gundam Blueprint II

Reward for completing True Mission: The Descending Sword (Directory 2)

Gouf Custom

Gundam Series: Mobile Suit Gundam: The 08th MS Team

Requirements:

3x Gouf Custom Blueprint I

Dropped by the first boss of Break Mission: The Shuddering Mountain, Part I (Directory 2). The boss must be defeated before he retreats

Seravee Gundam

Gundam Series: Mobile Suit Gundam 00

Requirements:

2x Seravee Gundam Blueprint I

Dropped by the second boss of Break Mission: The Shuddering Mountain, Part I (Directory 2)

Dropped by the first boss of True Mission: I Can Hear a Song (Directory 2)

Gundam Exia Repair II

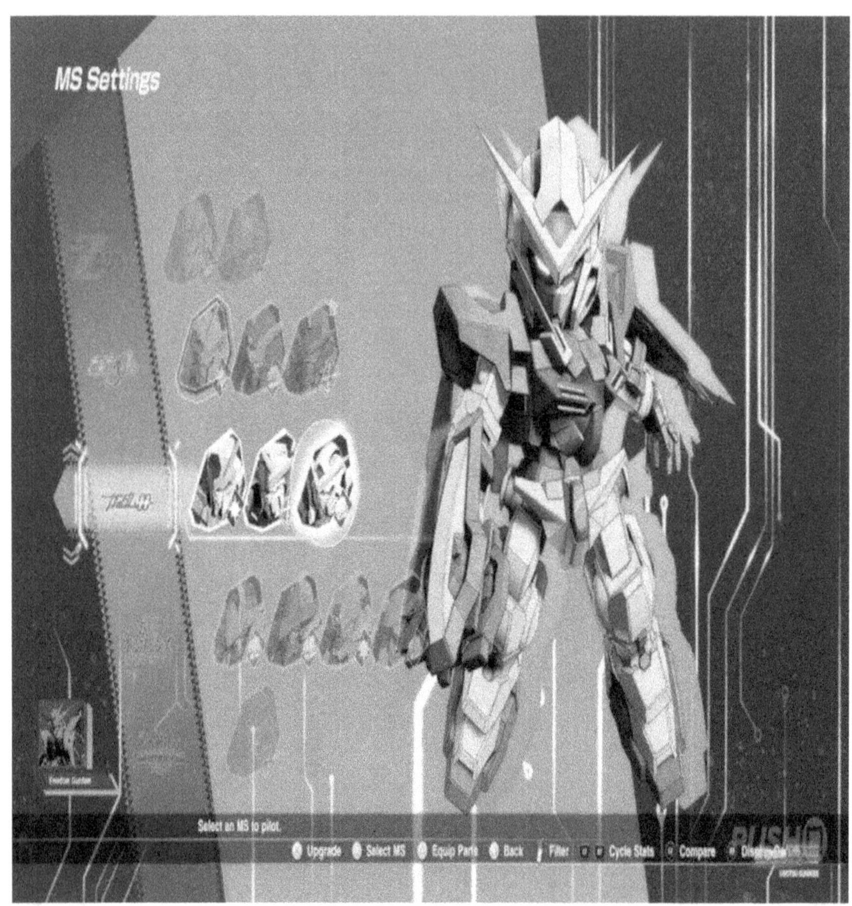

Gundam Series: Mobile Suit Gundam 00

Requirements:

3x Gundam Exia Repair II Blueprint I

Dropped by the first and second boss in True Mission: I Can Hear a Song (Directory 2)

2x Gundam Exia II Repair Blueprint II

Dropped by the boss of Irregular Mission: Trails to the Land (Directory 2)

Gundam Ez8

Gundam Series: Mobile Suit Gundam: The 08th MS Team

Requirements:

3x Gundam Ez8 Blueprint I

Dropped by the boss of True Mission: The Shuddering Mountain, Part II (Directory 2)

Reginlaze Julia (Final Battle)

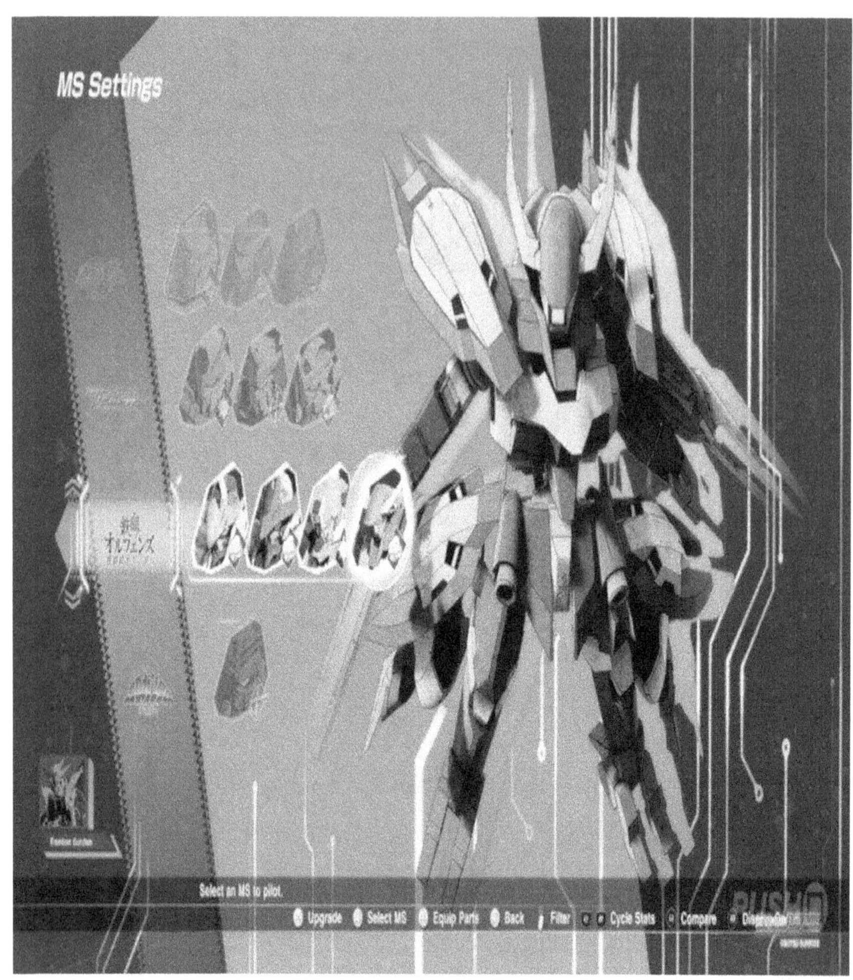

Gundam Series: Mobile Suit Gundam: Iron Blooded Orphans

Requirements:

1x Reginlaze Julia (Final Battle) Blueprint I

Dropped by the boss of Break Mission: Say it Ain't So, Bernie! (Directory 3)

Full Armor Gundam

Gundam Series: Mobile Suit Gundam Thunderbolt

Requirements:

4x Full Armor Gundam Blueprint I

Dropped by the boss of Break Mission: Battle of Solomon (Directory 3)

3x Full Armor Gundam Blueprint II

TBC

Kampfer

Gundam Series: Mobile Suit Gundam 0080: War in the Pocket

Requirements:

3x Kampfer Blueprint I

Dropped by the boss of True Mission: Over the River and Through the Woods (Directory 3)

Gundam "Alex"

Gundam Series: Mobile Suit Gundam 0080: War in the Pocket

Requirements:

3x Alex Blueprint I

Dropped by the boss of True Mission: Over the River and Through the Woods (Directory 3)

Z'Gok (Char)

Gundam Series: Mobile Suit Gundam

Requirements:

2x Z'Gok (Char) Blueprint I

Dropped by the boss of Break Mission: The Final Light (Directory 4)

1x Z'Gok (Char) Blueprint II

TBC

Providence Gundam

Gundam Series: Mobile Suit Gundam SEED

Requirements:

3x Providence Gundam Blueprint I

Dropped by the boss of Break Mission: Space Fortress: A Baoa Qu (Directory 4)

4x Providence Gundam Blueprint II

Dropped by the boss of True Mission: To a Future that Never Ends (Directory 4)

Zeta Gundam

Gundam Series: Mobile Suit Zeta Gundam

Requirements:

3x Zeta Gundam Blueprint I

Complete Break Mission: Space Fortress: A Baoa Qu after taking the West path through the level (Directory 4)

Dropped by the boss of True Mission: Forever Four (Directory 4)

3x Zeta Gundam Blueprint II

Dropped by the boss of True Mission: Forever Four (Directory 4)

Gundam Barbatos Lupus

Gundam Series: Mobile Suit Gundam: Iron Blooded Orphans

Requirements:

3x Gundam Barbatos Lupus Blueprint I

Complete Break Mission: Space Fortress: A Baoa Qu after taking the East path through the level (Directory 4)

Dropped by the boss of True Mission: Hunger of Angels (Directory 3)

4x Gundam Barbatos Lupus Blueprint II

Dropped by the boss of True Mission: Hunger of Angles (Directory 3)

Hyaku Shiki

Gundam Series: Mobile Suit Zeta Gundam

Requirements:

1x Hyaku Shiki Blueprint I

Complete Break Mission: Storm Over Kilimanjaro (Directory 4)

Gundam Gusion Rebake Full City

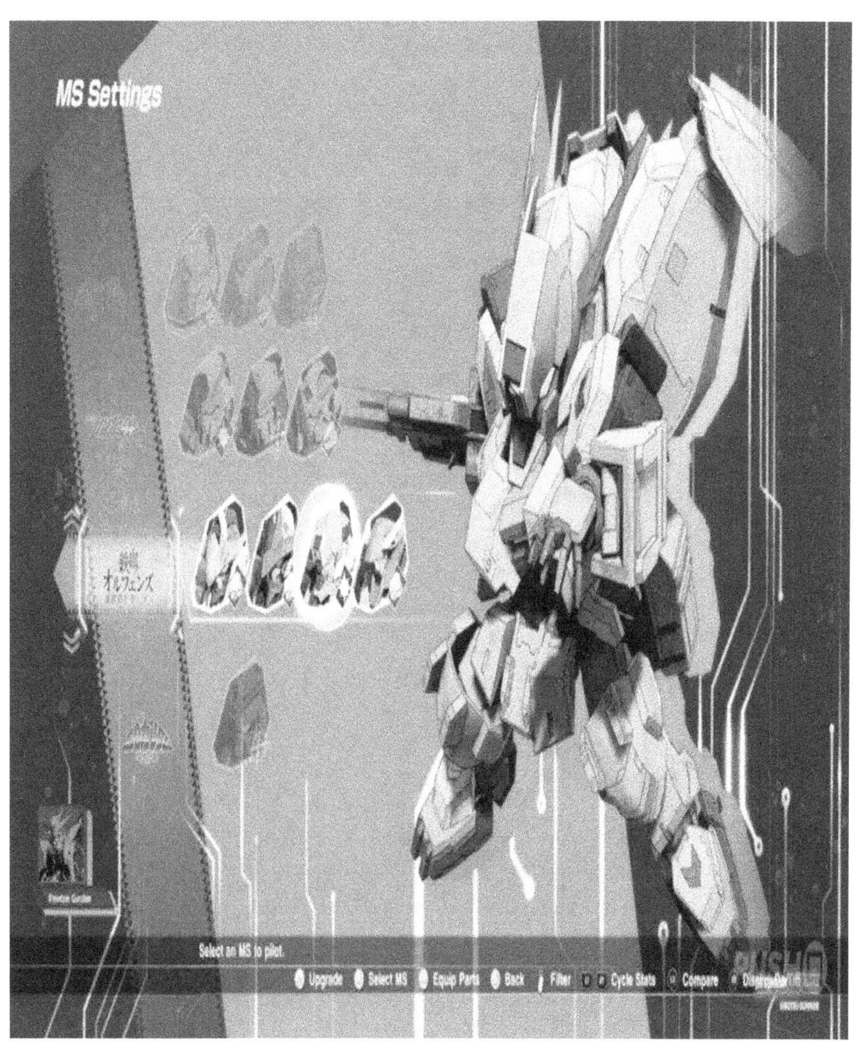

Gundam Series: Mobile Suit Gundam: Iron Blooded Orphans

Requirements:

3x Gundam Gusion Rebake Full City Blueprint I

Complete Break Mission: Storm Over Kilimanjaro (Directory 4)

Justice Gundam

Gundam Series: Mobile Suit Gundam SEED

Requirements:

3x Justice Gundam Blueprint I

Complete True Mission: To a Future that Never Ends (Directory 4)

Gelgoog (Char)

Gundam Series: Mobile Suit Gundam

Requirements:

1x Gelgoog (Char) Blueprint I

Dropped by the first boss in True Mission: Encounters in Space (Directory 4)

1x Gelgoog (Char) Blueprint II

TBC

Gundam GP03S Stamen

Gundam Series: Mobile Suit Gundam 0083: Stardust Memory

Requirements:

3x Gundam GP03S Stamen Blueprint I

Dropped by the first boss in True Mission: Encounters in Space (Directory 4)

3x Gundam GP03S Stamen Blueprint II

TBC

Gundam F91

Gundam Series: Mobile Suit Gundam F91

Requirements:

5x Gundam F91 Blueprint I

Dropped by the boss of Irregular Mission: Wings to the Heavens (Directory 4)

3x Gundam F91 Blueprint II

SD Gundam Battle Alliance Review

Every now and then, a licensed anime title comes along that goes above and beyond your expectations. Not just an ode to the source material, but a finely crafted video game as well. That's SD Gundam Battle Alliance in a nutshell — an almost dangerously addictive action RPG that franchise fans will likely adore. Despite its cutesy, chibi-style appearance, this is one of the best and most faithful Gundam adaptations that PlayStation has seen in a long, long time.

Battle Alliance is a mashup of nearly every Gundam series going. It all takes place within a kind of virtual reality, where Gundam histories are recorded and stored. Playing as a faceless VR pilot, you end up trapped within the database, and together with Juno, your operations officer, and an especially chipper AI nicknamed Sakura, you're tasked with repairing clusters of corrupt data to ensure your freedom. Naturally, this means taking to battlefields across loads of different timelines, and putting thousands of cute mobile suits to the sword.

The story itself is largely forgettable and full of shaky sci-fi terms that don't really mean anything, but it acts as a serviceable vehicle for ferrying you between missions. More than anything, it's the gameplay loop that'll keep you coming back — an excellent mix of bite-sized battles, tense boss fights, and the joys of collecting and upgrading an extensive catalogue of playable mobile suits.

If you're a Gundam enthusiast, there's a genuine thrill in seeing which suits and pilots pop up next.

The aforementioned database corruption means that characters from far away realities can collide, leading to some especially cool crossover encounters.

Once they've been sent back to their own world, you unlock 'True' missions where you get to re-enact the canon scenario, accurate character dialogue and all.

The game even recreates key scenes from the various anime, further solidifying the feeling that this is a real labour of love from developer Artdink.

Moving on, combat is straightforward on a surface level. Each suit has a one-button melee combo, a heavy attack, and three special moves, usually reliant of ranged weaponry that operates on a cooldown.

A super attack is also an option once your gauge is maxed, and your defensive techniques consist of blocking and dodging.

Again, straightforward stuff, but there's a decent degree of depth to be found in perfectly timed blocks, dodges, and aerial combos. It's all snappy, easy to understand, and fun to utilise.

Taking on standard enemy suits is fairly simple, and perhaps a tad repetitive after you've invested tens of hours into the game, but cleaving an opponent in half with your trusty beam saber is always satisfying.

Boss battles, on the other hand, can be much more demanding.

The vast majority of missions end with a big fight against a deadly opponent, decked out with a beefy health bar and uninterruptable attacks.

These encounters are all about patience; waiting for the right moment to launch a counterattack. Think of it like Monster Hunter, but with funny little robots.

These duels are particularly difficult if you're alone, so you'll be thankful for your AI companions. You can take two allies with you on missions, and although your buddies can be a bit stupid, bosses can only focus on one target at a time. As such, it feels like boss encounters are designed around team-based play, with mechanics like increased back damage being crucial in chipping away at the huge health pools. What's more, you and your AI pals can revive each other, should things go pear-shaped.

Some boss battles, specifically against giant mobile armour enemies, are especially well designed in that they encourage you to properly explore your suit's capabilities. Perfect blocks and dodges can become a necessity when the going gets tough, and mastering a tricky fight feels like a reward in itself.

So combat's generally great fun and unlocking all kinds of mobile suits is a blast, but it should be stressed that there is a grind here. Individual missions only tend to last between five and ten minutes, but you'll find yourself replaying them in order to both unlock and level up your shiny new suits. The excitement of finally getting your hands on a cool Gundam is often tempered by the realisation that yet more grinding will be required, if only to bring it up to speed with your current suit of choice.

Co-op can certainly help alleviate any tedium, though. You can create your own online lobby or jump into an open one through the main menu, and any progress that you make carries back into your single-player campaign — including the unlocking of later missions. As you can imagine, co-op play reveals new strategies, and if you're teaming up with a couple of pilots who know what they're doing, it's easily the most efficient way to get things done.

But if you're running into trouble and you don't want to hop online, you can always make use of the game's easy mode, which makes missions much less demanding. The only downside is that you won't find as many parts — equippable stat-boosting items — during your excursions. In any case, a great place to start if you want to ease yourself into the experience.

We're making this same point again, but SD Gundam Battle Alliance really is a feast for franchise aficionados. Graphically, the game isn't at all impressive, but the suits, in all of their chibi glory, have been perfectly modelled. Likewise, the property's many trademark sound effects are here, and the soundtrack is stuffed with music from the shows alongside banging electronic remixes. For wannabe Newtypes, it's nothing short of a joy.

Conclusion

SD Gundam Battle Alliance is one of the best Gundam games in what feels like an age. A highly addictive gameplay loop carries the experience, consisting of bite-sized missions, snappy combat, and the thrill of discovering and unlocking new mobile suits. The grind can feel a bit aggressive at times, and the story's vapid, but there's a clear love for all things Gundam here.